Bibliographic information published by the German National Library:

The German National Library lists this publication in the National Bibliography; detailed bibliographic data are available on the Internet at http://dnb.dnb.de .

Imprint:

Print and binding: Books on Demand GmbH, Norderstedt Germany
ISBN: 9783668319790

This book at GRIN:

http://www.grin.com/en/e-book/341767/geo-biologically-important-oil-indicating-microorganisms-of-sub-saharan

Debajyoti Bose

Geo-biologically Important Oil Indicating Microorganisms of Sub-Saharan region. Screening of Oil Indicating Microorganisms from Major Geo-Political Zones of Yobe State-Nigeria

GRIN Publishing

RESEARCH PROJECT

SCREENING OF OIL INDICATING MICROORGANISMS FROM MAJOR GEO-POLITICAL ZONES OF YOBE STATE-NIGERIA

READER AND HEAD
DEPARTMENT OF BIOLOGICAL SCIENCES
YOBE STATE UNIVERSITY
DAMATURU, YOBE STATE, NIGERIA

ACKNOWLEDGEMENT

Author highly acknowledges Yobe State University, Damaturu, Yobe State, Nigeria for providing laboratory facilities for this research work.

Dr.Debajyoti Bose
Reader & Head
Department of Biological Sciences
Yobe State University
Damaturu, Yobe State, Nigeria

ABSTRACT

Ancient rocks and natural oils are rich with abundant and hidden information of ancient organic matters responsible for earth formation, including molecular and isotopic signatures of the organisms that existed at the time the organic matter was formed. Considering the above facts of astrobiology, research study was carried out on screening of oil indicating microorganisms from oil contaminated sites of major geo-political zones of Yobe State, Nigeria including Damaturu, Gashua, Potiskum and Nguru local Government areas. Serial dilution was performed for all soil samples collected from the oil contaminated sites of mentioned local government areas followed by culture isolation and identification techniques for bacterial samples only. Nutrient agar was used for the isolation of the bacteria from the soil samples and mixed cultures were obtained. These bacterial cultures were inoculated in petroleum in order to see their oil utilizing capability. Two types of culture showed positive results among all isolated colonies. Rhizoid shaped gram positive rods isolated from soils of Damaturu, Gashua, Potiskum and Nguru & yellow pigmented gram negative bacterial colony isolated from soils of Potiskum and Nguru showed oil utilizing ability by positive growth in the refined crude oil (Petroleum, Diesel, etc.). The oil utilizing ability of those specific bacteria shows their potentiality in detecting the presence of oil and acting as potential oil indicators in Sub-Saharan region of African Continent caring a long history of bio-geo sciences. Findings from the present research work may create an opening for encouraging the geo-biological potentials of West African Countries in astrobiological aspects.

Key words: Geo-political zones, Oil indictor, bacteria, refined crude oil, bio-geo sciences, astrobiology.

TABLE OF CONTENTS

CHAPTER ONE GENERAL INTRODUCTION

1.0. Introduction

Geo-microbial prospecting for hydrocarbons is an exploration method on the premise that the light gaseous hydrocarbons migrate upward from subsurface petroleum accumulations by diffusion and effusion, and are utilized by a variety of microorganisms present in the sub-soil ecosystem. The hydrocarbon oxidizing bacteria exclusively use these gases as carbon source for their metabolic activities and growth. These bacteria are mostly found enriched in the shallow soils/sediments above hydrocarbon-bearing structures and can differentiate between hydrocarbon prospective and non-prospective areas.

1.1. Background of the Study

A microbial prospecting technique for hydrocarbon research and exploration has been in use for the last five decades. The isolation and enumeration of specific alkane-oxidizing bacteria are used as indirect petroleum prospecting method. Microbial anomalies have been proved to be reliable indicators of oil and gas in the sub-surface (Pareja, 1994). The direct and positive relationships between the microbial population and the hydrocarbon concentration in the soils have been observed in various producing reservoirs worldwide (Miller, 1976; Sealy, 1974; Wagner et al., 2002).

The microbial survey was first proposed and applied in the U.S.S.R. Early use of bacterial soil flora as a means of detecting gas as a means of detecting gas seepages was a development stemming from oil gas surveys performed in the U.S.S.R (Mogilevskii, 1959). The initial microbial investigations of Mogilevskii and his associates in the field of petroleum prospecting incited the interest of petroleum geochemist worldwide.

It has been shown that out of 20 microbial anomalies, 16 were proved by successful drilling. Sealy(1974a) in U.S.A, carried out microbial prospecting surveys and showed a positive correlation of 85.7%. Miller(1976) reported microbial survey carried out in the oil fields of the U.S.A, in which the microbial activity profile indicated a good contrast between oil field and nearby dry area.Beghtel (1987) predicted hydrocarbon potential of 18 wildcat wells in the Kansas; out of which, 13 have proved to be commercial produces oil and gas. As a result, microbial methods for detecting petroleum gas in the soils and waters have been tested in Europe, U.S.A and elsewhere. A sustained effort for the presence of efficiently utilizing microbiological prospecting methods had been underway since 1938 including independent surveys, surveys coordinated with geochemical and geophysical operation, microbiological tests of both soils and subsurface waters and analysis for methane oxidizing as well as higher hydrocarbon-oxidizing bacteria. However, oil and gas fields also build up micro-seepages at the subsurface soils, and these micro-seepages are detectable using a variety of analytical techniques that have been developed the basis for these new surface prospecting methods. The pioneers were Laubmeyer (1993) in Germany, Rosaire (1939) and Horvitz (1939) in the United States. Using methods such as extraction of adsorbed hydrocarbon gases from surface samples, they documented a correlation between higher hydrocarbon concentrations and oil and gas fields. At almost at the same time, the microbiologists Mogilevskii (1938, 1940) in the U.S.S.R. and Taggart (1941) and Blau (1942) in the United States described the use of hydrocarbon-oxidizing bacteria, when measured in surface soil samples, as an indicator for oil and gas fields in the deeper subsurface. In the 1950s and early 1960s, many relevant publications came from the United States (Updegraff et al., 1954; Davis, 1956). Several microbiological methods for detecting the distribution and activity of hydrocarbon-oxidizing bacteria were developed, such as enumeration of cell content in soil samples, measuring gas-consumption rates and radioautography. The field trials conducted by Sealy (1974) in U.S.A., using microbiological techniques, showed that out of 89 locations tested in West Texas, Wildcats predicted as productive and non-productive correlation of 54% and 92% respectively.

Microbial prospecting survey has been widely used in Germany since 1961 and a total of 17 oil and gas fields were identified. The success rate of Microbial Prospecting for Oil and Gas (MPOG) method has been reported to be 90%.

1.2. Statement of the Problem

Most exploration depends on highly sophisticated technology to detect and determine the extent of hydrocarbon deposits using exploration geophysics. Areas thought to contain hydrocarbons are initially subjected to a gravity survey, magnetic survey, passive seismic reflection or regional seismic reflection surveys to detect large scale features of the subsurface geology. Features of interest (known as leads) are subjected to more detailed seismic surveys which work on the principle of the time it takes for reflected sound waves to travel through matter (rock) of varying densities and using the process of depth conversion to create a profile of the substructure. Finally, where a prospect has been identified and evaluated and passes the oil company's selection criteria an exploration well is drilled in attempt to conclusively determine the presence or absence of oil and gas.

This exploration technique is an expensive, high risk operation. Offshore and remote area exploration is generally only undertaken by very large corporations or national governments.

In 1956, Shell British petroleum discovered crude oil at village Oloibiri in Bayelsa state in the Niger Delta within Nigeria (Anifowose, 2008; Onuhua, 2008) and since then oil exploitation and exploration have been on-going for several decades. However, that area has seen increasing militant actions by locals unhappy with both their exploitation and the environmental destruction caused by the exploitation. Recently, it was reported according to some geological observations, that the northern part of Nigeria has massive quantities of oil and there is need for modern technology to explore this oil in the rich areas. Nevertheless, as the northern part of Nigeria is in the savannah region, the geophysical data is hardly obtained.

1.3. Research Questions

i) Why do microorganisms oxidize/utilize light gaseous hydrocarbons in contaminated soil?

ii) How do hydrocarbon-oxidizing bacteria have positive relationship with hydrocarbon concentration in the soil?

iii) Why are hydrocarbon-oxidizing bacteria used as indicators for finding oil and gas reservoirs.

1.4. Aim and Objectives (Purpose) of the Study

The aim of this study is to isolate oil degrading microorganisms from soils of major areas of Yobe State.The objectives of this study are stated below:

i) To provide efficient, less expensive method of oil prospecting in area, thus reducing the risk of operation.

ii) To provide method that is helpful in the quick evaluation of hydrocarbon prospects and generation of exploratory leads.

iii) To provide a means of identification for prospective oil and oil and gas in areas where no geophysical data is available or where such investigation is difficult.

iv) To provide a means that can be integrated with geological, geophysical and geochemical methods to prioritize drilling locations thereby reducing drilling risks and achieving higher success in petroleum exploration.

1.5. Significance of the Study

This study will be of great importance to oil explorers since it will help to provide a cost effective, low-risk means of oil prospecting in an area. It will also help them in identification of oil and gas in areas where no geographical data is available or where such investigation is difficult. The study will also help in the identification/discovery of oil in rich areas. It will serve as a reference point for future researchers who will want to research more on the topic.

1.6. Scope and Limitation of the Study

This research tends to detect the presence of oil degrading microorganisms in contaminated sites in Yobe State. However, the research will be restricted to four selected towns in the major zones of Yobe State. The towns are Damaturu, Potiskum, Gashua and Nguru.

1.7. Operational Definition of Terms

i. **Exploration:** examination of something in order to find out about it.

ii. **Geochemistry:** derived from Geology and Chemistry describing the nature and appearance of matter together with its chemical composition.

iii. **Geo-microbial prospecting:** derived from geology and microbes, describing the use of microorganisms for prospecting oil.

iv. **Geophysics:** scientific study of the earth's atmosphere, oceans and climate.

v. **Hydrocarbons:** chemicals made of hydrogen and carbon only found in petroleum and gas.

vi. **Hydrocarbon-oxidizing bacteria:** the bacteria responsible for utilizing/oxidizing hydrocarbons.

vii. **Isolation of microbes:** the growing a pure culture of (a microbe), usually as individual colonies on a solid medium. Or, it is the separation of pure strain of a microorganism or virus.

viii. **Oil seepage:** the process by which oil flows slowly and in small quantities through the soil.

CHAPTER TWO LITERATURE REVIEW

2.0. Introduction

Bacteria are ubiquitous in distribution and their exceptionally high adaptability to grow on different nutrient sources form the basis of microbial prospecting (Rasheed et al, 2013). Several investigators have used bacteria that degrade hydrocarbons used as indicators for finding oil and gas reservoirs.

Sealy (1974) in U.S.A, carried out microbial prospecting surveys and showed a positive correlation of 85.7%. Miller (1976) reported microbial survey carried out in the oil fields of the U.S.A, in which the microbial activity profile indicated a good contrast between oil fields and nearby dry area. Beghtel (1987) predicted hydrocarbon potential of 18 wildcat wells in the Kansas; out of which, 13 have proved to be commercial producers of oil and gas.

Rasheed et al., (2013) carried out geo-microbial prospecting method for hydrocarbon exporation in Vengannapalli village, Cuddapah Basin India, with a view to investigate whether the leakage of natural gas is associated with some geo-microbial anomalies that can establish an upward migration of hudrocarbons from deep subsurface source, where they report high bacterial counts for methane (5.4×10^5cfu/g) and propane-oxidizing bacteria (4.6×10^4cfu/g) in soil samples collected from Vengannapalli village. The adsorbed light gaseous hydrocarbon analysis shows the presence of moderate to low concentrations methane, ethane, propane and butane in the soil samples. Carbon isotope analysis of few soil samples show enriched δ^{13} C, values of methane ranging from -35.0 to 10.6‰, indicative of methane oxidation. Geo-microbial prospecting method coupled with adsorbed soil gas and carbon isotope studies, suggest that hydrocarbon micro-seepage of subsurface origin is present in the area and indicate that the area is worth visiting for conventional petroleum exploration. Similar technique was used by Rasheed et al., in Deccan Syncline, Maharashta, India. Subsurface soil samples were collected in an interval of 5×5 km, and analysed for propane oxidizing bacterial concentration. The propane oxidizing bacterial count in the soil samples of the studied area ranged from 1 × 10^2 to 6.7 × 10^5 cfu/gmof soil sample. Two microbial blooms of high concentration of propane oxidizing bacteria were identified and mapped in the study area.

In a similar research by Jahir A. K. and Sayed A. R. (2011), on 'Isolation and characterization of micro-organisms from oil contaminated sites in Lucknow, the researchers collected soil samples from two different oil contaminated sites (used oil) and isolated four used oil degrading bacteria. One isolate labelled by the researchers as MJH101 showed maximum oil degradation abilities. Only one ml was recovered out of 25ml after seven days of incubation, and there was a decrease in width of oil layer from 6mm to 1mm.

Gupte, A. and Sonawdekar, S. (2015), also isolated oil degrading bacteria from oil contaminated sites in and around Navi, Mumbai, India. Nineteen different isolates were obtained from ten oil polluted areas. The oil degrading capacity of these isolates was evaluated using Emulsification Index and Gravimetric Analysis. The oil emulsification index was found to be in the range of 25 to 54%. Amongst the isolates, culture 1 showed best emulsification index and hence was used for further study. Oil degrading capacity of this isolate was found to be 41%. GC-MS study showed the conversion of heinocosane to hexadecane indicating that the isolate is *Bacillus cereus*. However, their research is on the use of these microorganisms as bioremediation agents.

2.1. Microbial Techniques for Hydrocarbon Exploration

Microbial technique for hydrocarbon exploration is a method based on the seepage of light gaseous hydrocarbons from oil and gas reservoirs towards the surface and their utilization by hydrocarbon-oxidization bacteria (Rasheed et al., 2008). The methane, ethane, propane, and butane-oxidizing bacteria exclusively use these gases as carbon source for their metabolic activities and growth. These bacteria are mostly found enriched in the shallow soils/sediments above hydrocarbon bearing structures and can differentiate between hydrocarbon prospective and non-prospective areas (Tucker and Hitzman, 1994). The detection of various groups of methane, ethane, propane or butane oxidizing bacteria, in the surface soils or sediments helps to evaluate the prospects for hydrocarbon exploration. Microbial anomalies have been proved to be reliable indicators of oil and gas in the subsurface (Pareja, 1994). The direct and positive relationships between the microbial population and the hydrocarbon concentrations in the soils have been observed in various producing reservoirs worldwide (Sealy 1974a, 1974b; Wagner et al., 2002).

The microbial prospecting method involves the collection of sub-soil samples from the study/survey area, packing of samples, preservation and storage of samples in pre-sterilized sample bags under aseptic and cold conditions till analysis followed by isolation and enumeration of hydrocarbon utilizing bacteria. The contour maps for population density of hydrocarbon oxidizing bacteria are drawn and the data is integrated with other geo-scientific and geophysical data to find the hydrocarbon prospectivity of the area (Rasheed et al., 2013).

2.2. Methane Oxidizing Bacteria

Methane oxidizing bacteria were the first type of bacteria studied to identify the location of petroleum accumulations. The presence of methane oxidizing bacteria in the soil, in the absence of cellulose oxidizing bacteria, has been interpreted as indicating the presence of methane exhalation from the sub-surface (Tedesco, 1995). Mogilevskii (1938) described the possibility of using methane oxidizing bacteria for gas exploration. Methane oxidizing bacteria have been isolated from soil as a means of prospecting from natural gas and/or oil deposits (Brisbane and Ladd, 1965). The methane oxidizing bacteria are usually predominant over gas fields as the gas reservoirs are commonly dominated by methane. Microbial Prospecting for Oil and Gas (MPOG) method establish the separate activities of methane oxidizing bacteria as gas indicators and those bacteria that oxidize only ethane and higher hydrocarbons as oil indicators, it is possible to differentiate between oil fields with and without free gas cap, and gas fields (Wagner et al., 2002). Whittenbury et al., (1970) reported the isolation of more than 100 strains of methane oxidizing bacteria. Hanson and Watternberg (1991) and Hanson and Hanson (1996) gave overviews of the ecology of methylotrophic bacteria and their role in the methane cycle. The methane oxidizing bacteria (methylotrophs) are *Methylococcus*, *Methylomonas*, *Methylobacter*, *Methylocytis*, *Methylosinus*, *Methylobacterium*e.t.c., (Hanson and Hanson 1996). However, methane oxidizing bacteria are known to be poor indicators in petroleum prospecting because methane can occur in the absence of petroleum deposits and moreover, some methane oxidizing bacteria are unable to oxidize other aliphatic hydrocarbons. Detection of ethane and longer chain hydrocarbon oxidizing bacteria on the other hand provides presumptive evidence for a hydrocarbon seep and an underlying petroleum reservoir (Davis, 1967). The principal advantage of using methane oxidizing bacteria for petroleum prospecting is the preponderance of methane in petroleum gas and it is the most mobile of petroleum hydrocarbons and this cannot be ignored (Rasheed et al., 2013).

2.2.1. Microbial Oxidation of Methane

Microbial oxidation of methane starts with an activation of methane by methane-oxygenase because only one atom of the hydrogen molecule is inserted into the methane molecule to provide methanol; and leads, in the presence of oxygen, to methanol and then to formaldehyde. Formaldehyde can be directly assimilated to produce biomass or oxidized to CO_2 for the generation of energy (Leadbetter and Foster 1958). Because of the high specialization of these bacteria, methane oxidizers can be isolated from all other bacteria. The successful detection of these specialists indicates methane occurrence in soil samples (Wagner et al., 2002).

2.3. Ethane, Propane and Butane Oxidizing Bacteria

The soviets were first to use microbial prospecting method. They established that certain bacteria specifically consume ethane, propane, or butane, but not methane. These hydrocarbons are assumed to be from petroleum migrating from depth and are not associated with generation in the soil (Tedesco 1995). The isolation and enumeration of specific C_{2+} alkane oxidizing bacteria has been used as indirect petroleum prospecting method (Davis, 1967; Wagner et al., 2002). The detection of bacteria that oxidize n-alkanes having chain lengths of 2 to 8 carbon atoms, without any adaptation period, indicates the existence of short-chain hydrocarbons in the investigated soil samples, and thus can indicate the presence of oil accumulations in the subsurface. In those areas in which both short-chain hydrocarbons and methane are detected in this manner, one can assume a thermogenic gas with significant quantities of short-chain hydrocarbons ethane, propane, and butane can be used by a large number of bacteria (*Mycobacteria,Flacobacteria*, *Nocardia*, and *Pseudomonas*). The microbial anomalies have been proven to be reliable indicators of oil and gas in the subsurface (Beghtel et al., 1987; Lopez et al., 1993; Tucker and Hitzman, 1994). The microbial prospecting method has been used to prioritize drilling locations and to evaluate hydrocarbon prospects of an area (Pareja, 1994), thus reducing drilling risks and achieving higher success ratio in petroleum exploration.

2.3.1. Microbial Oxidation of Ethane, Propane and N-Butane

The microbial oxidation of ethane, propane and n-butane proceeds stepwise to alcohols, then to aldehydes and finally to acetate, which can be assimilated into cell material. The number of species of bacteria capable of using such hydrocarbons in this process increases with the expansion of the chain length of the alkanes. The degradation of the alkane occurs by terminal oxidation by means of mono-oxygenase and by β-oxidation to acetyl-CoA, which is the initial substance in several biochemical reactions (Atlas, 1984).

2.4. Sample Collection Method.

The geo-microbial surveys of prospecting involve collection of suitable samples from sub-soil horizon and detection of specific micro flora in the samples. Sampling is important since the validity of the test results depend largely upon the manner in which the samples are taken. The samples are collected using a hollow metal pipe by manual hammering. The soil samples of about 100gm each are collected in pre-sterilized bags under aseptic conditions from a depth of about 0.5 to 1m and preferably on a grid pattern with a spacing of 200m (Davis, 1967). However, it may be less depending on the type of the research. The samples are immediately stored at 2 to 4°C and subsequently transported to the laboratory and are stored under cryonogenic conditions till analysis. Soil samples should not be stored for a long time, and samples should be analysed as early as possible after collection. Sampling should not be done in disturbed or excavated areas, soils contaminated with hydrocarbons, chemicals or animal wastes, swamps and areas under water shed. During collection of soil samples, rocks, coarse materials, plant residues, and animal debris have to be excluded (Rasheed et al., 2013).

2.5. Isolation of Hydrocarbon Oxidizing Bacteria

The specific bacterial populations are measured by standard microbiological screening techniques for hydrocarbon-indicating microorganisms. A selective growth media is used which cultures only microorganisms capable of utilizing light hydrocarbons. Various methods such as bacterial pellicle formation, gas uptake, soil dilution and plating methods have been widely used. Other techniques can determine the presence of living bacterial cells in soil samples. The determination of the number of colony-forming units (CFUs) in solid feeding media and the most-probable number (MPN) procedure in nutrient solutions can be applied (Rasheed et al., 2013).

2.6. Molecular Biology Techniques

Currently, molecular biology technique has achieved great development in studies of soil samples. Development of molecular biology methods for microbial prospecting for oil and gas by applying culture independent techniques will improve the accuracy rate of microbial prospecting for oil and gas exploration (Fan Zhang et al., 2010). The most-probable-number (MPN) procedure has traditionally been applied to determine the numbers of colony forming units (CFUs) in soil samples. The real-time polymerase chain reaction (RT-PCR) is now being widely used to detect and quantify various target microorganisms without experimental cultivation (Dionisi et al., 2003; Skorhus et al., 2004; He et al., 2007). Molecular techniques related with 16s ribosomal D.N.A (rD.N.A) has been proven effective as a basis for understanding the microbial diversity in environmental communities. The cloning and sequencing of 16s rD.N.A is sufficient for the identification of the microorganisms present in a given habitat and for the discovery of previously unknown diversity (Hugenholtz et al., 2008). These techniques were also applied to investigate microbial communities in the formation water of the produced water of oil fields (Kaster et al. 2009; Lysnes et al. 2009). Characterization of microbial communities involved in short-chain alkane metabolism, namely methane, ethane and propane, in soil samples from a petroliferous soils through clone libraries of the 16S rRNA gene of the Domains Bacteria and Archaea and the catabolic gene coding for the soluble di-iron monooxygenase (SDIMO) enzyme alpha subunit. Further studies on the occurrence and diversity of SDIMO genes in soil, as well as the improvement of primer sets to be applied in real-time PCR, are necessary in order to overcome the obstacle of the low abundance of catabolic genes in natural environments and enable their quantification from complex genetic backgrounds (Paula, et al., 2011). Immunological or DNA probes for gaseous hydrocarbon utilizing bacteria; Researchers are on for developing immunological probe or DNA probe which will rapidly identify the specific hydrocarbon oxidizers from the survey area. Immunological probes are based on the technique of preparing monoclonal antibodies as probes. In DNA probes, a selective small piece of DNA either from plasmid or from chromosomal segment serves as a probe. This segment is coded with the gene sequence, responsible for specific hydrocarbon gases oxidation. DNA hybridization with homologous st rain gives the detection of hydrocarbon oxidizers. Detection of Biomarkers – Diploterol; the hapanoiddiploterol is a known product of various aerobic bacteria that is particularly abundant in methanotrophic bacteria. The methanotrophs are in fact the source of diploterol is indicated by the compound light isotopic composition or around -60 per mil. Moreover 12C enriched hydrocarbon derivatives of diploterol were found in other studies of sediments in seep environments. This finding makes it a promising developmental technique for bio-prospecting of hydrocarbons. The above methods are the future thrust areas in bio-processing and will enhance the efficacy of the technique.

CHAPTER THREE MATERIALS AND METHODS

3.1. Collection of Soil Samples

Soil samples were collected from four towns in Yobe and all from filling stations where the soil is contaminated. These towns are Damaturu, Potiskum, Gashua and Nguru. From each site, the surface soil as well as the soil at 10cm below the surface were collected respectively and packed in sterile containers. The collected samples were immediately transferred to the laboratory. The table below shows the area of sample collection and the temperature of the area at the time of collection of sample.

Table 1: Areas of soil sample collection and the temperature of areas

s/n	Area of soil sample collection	Temperature of the area
1	Damaturu	$37 \pm 1^{o}C$
2	Potiskum	$37 \pm 1^{o}C$
3	Nguru	$38 \pm 1^{o}C$
4	Gashua	$40^{o}C \pm 1^{o}C$

3.2. Isolation of Bacteria from Soil Samples

For isolation of bacteria from soil sample the following microbiological techniques were used.

3.2.1. Serial Dilution

The soil samples were diluted several times to reduce the microbial population sufficiently to obtain separate colonies on petri plates. Serial dilutions were performed for all the soil samples where in the soil samples were diluted from 10^{-1} to 10^{-5} dilutions respectively.

3.2.2. Media Preparation and Sterilization

The nutrient media used for isolation of bacteria from the soil sample in this research was Nutrient Agar. 14g of the Nutrient Agar powder was dissolved in 500ml distilled water in a sterile conical flask. The media was then autoclaved at $121^{o}C$ for 15 minutes at 15 psi. It was allowed to cool for a couple of minutes. 25 ml of the cooled media was transferred to 16 different sterilized plates respectively. The media was allowed to solidify in the plates.

3.2.3. Spread Plate Method

1ml of diluted microbial mixture was transferred to the nutrient agar plate and spread over the surface with a sterile bent-glass rod. The plates were incubated at $37^{o}C$ for 48 hours. The dilution factors used for the inoculation were 10^{-3} and 10^{-4} respectively.

3.3. Gram's staining

After 48 hours of incubation, mixed cultures were obtained and in order to confirm whether the bacteria are gram positive or gram negative, the gram's staining was performed.

Gram's Staining Procedure:

i. A loop full of culture was picked with a sterile inoculating loop and was smeared on glass slide.

ii. The smear was heat fixed and air dried.

iii. The smear was first flooded with the primary stain, crystal violet in this case, for about 1 minute.

iv. The smear was rinsed with water for about 1 minute to remove excess crystal violet dye and then flooded with dilute solution of iodine, called gram's iodine for about 1 minute.

v. The stained smear was rinsed again, and then 95% ethanol was briefly added (about 30 seconds) to decolorize gram negative bacteria.

vi. A counterstain was then applied to impart a contrasting colour to the now colourless gram negative bacteria. For this purpose, the pink-dye safranin was used.

vii. The slide was then rinsed with water for the last time. It was allowed for some minutes to dry. Oil immersion was added on the slide and the slide was observed under 100x objective lens of a microscope.

3.4. Application of Oil Utilizing Bacteria

The following experiment was carried out to determine how oil utilizing bacteria can be used as oil indicators:

i. 5ml of petroleum was added to a sterile test tube.

ii. A loop full of culture from the mixed cultures was transferred to the test tube containing the petroleum.

iii. The inoculated test tube was then incubated at 37°C for 48 hours.

iv. One test tube containing the 5ml petroleum was not inoculated, but was incubated to serve as control.

The above steps were performed for all the different types of cultures isolated from the soils.

CHAPTER FOUR RESULTS AND DISCUSSION

4.1. Results

At the end of each test conducted, observations were recorded, such as colony characterization; gram's staining, as well as the application of oil degrading bacteria.

4.1.1. Colony Characterization

After 48 hours of incubation, colony characterization was done for the mixed cultures obtained.

Table 2: Colony characteristics of culture obtained from Damaturu soil sample at 10^{-3} and 10^{-4} dilutions

s/n	Colony characteristics	Surface soil		10cm depth soil	
		10^{-3}	10^{-4}	10^{-3}	10^{-4}
1	No. of colonies	59	70	51	55
2	Shape	11 irregular, 48 spherical	2 rhizoid, 6 irregular, 62 spherical	6 rhizoid, 19 spherical, 30 irregular	12 irregular, 39 spherical
3	Elevation	34 raised, 25 flat	45 raised 25 flat	22 raised, 29 flat	8 raised, 47 flat
4	Appearance	Opaque	Opaque	Opaque	Opaque
5	Texture	58 smooth, 1 rough	Smooth	Smooth	Smooth
6	Pigmentation	1 red, 1 yellow	1 pink	6 yellow, 2 pink	3 yellow, 2 pink

Table 3: Colony characteristics of culture obtained from Potiskum soil sample at 10^{-3} and 10^{-4} dilutions.

s/n	Colony characteristics	Surface soil		10cm depth soil	
		10^{-3}	10^{-4}	10^{-3}	10^{-4}
1	No. of colonies	60	37	34	16
2	Shape	3 irregular, 57 spherical	4 irregular, 33 spherical	5 irregular, 29 spherical	1 rhizoid, 5 irregular, 10 spherical
3	Elevation	Raised	Raised	Raised	10 raised, 6 flat
4	Appearance	Opaque	Opaque	Opaque	Opaque
5	Texture	1 rough, 59 smooth	1 rough, 36 smooth	1 rough, 33 smooth	1 rough, 13 smooth
6	Pigmentation	1pink,3yellow, 1 white	3 yellow, 1 pink, 1 white	5 yellow, 3 pink	4 yellow, 1 red, 1 pink

Table 4: Colony characteristics of culture obtained from Nguru soil sample at 10^{-3} and 10^{-4} dilutions.

s/n	Colony characteristics	Surface soil		10cm depth soil	
		10^{-3}	10^{-4}	10^{-3}	10^{-4}
1	No. of colonies	54	47	59	22
2	Shape	1 rhizoid, 4 irregular, 49 spherical	1 rhizoid, 2 irregular, 44 spherical	1 rhizoid, 3 irregular, 55 spherical	3 irregular, 19 spherical
3	Elevation	51 raised, 3 flat	Raised	1 filamentous, 2 flat, 57 raised	Raised
4	Appearance	Opaque	Opaque	Opaque	opaque
5	Texture	3 rough, 51 spherical	1 rough, 46 smooth	Smooth	1 rough, 21 smooth
6	Pigmentation	3 white, 13 yellow	1 yellow	3 yellow	1 white

Table 5: Colony characteristics of culture obtained from Gashua soil sample at 10^{-3} and 10^{-4} dilutions

s/n	Colony characteristics	Surface soil		10cm depth soil	
		10^{-3}	10^{-4}	10^{-3}	10^{-4}
1	No. of colonies	54	23	27	19
2	Shape	3 irregular, 51 spherical	4 irregular, 19 spherical	2 irregular, 25 spherical	1 rhizoid, 4 irregular, 14 spherical
3	Elevation	50 raised, 4 flat	18 raised, 5 flat	20 raised, 7 flat	10 raised, 9 flat
4	Appearance	opaque	Opaque	Opaque	Opaque
5	Texture	smooth	Smooth	Smooth	smooth
6	Pigmentation	3 pink, 3 yellow	1 pink, 2 red	3 pink, 1 yellow	1 pink, 1 red

4.1.2. Observations for Gram's Staining

Table 6: Gram's staining result for cultures obtained from Damaturu soil sample

s/n	Colony type	Surface soil	10cm depth soil
1	Spherical	Gram positive cocci	Gram positive cocci
2	Irregular	Gram positive cocci	Gram positive cocci
3	Rhizoid	Gram positive rod	Gram positive rod
4	Yellow pigmented	Gram positive cocci	Gram positive cocci
5	Pink pigmented	Gram positive cocci	Gram positive cocci
6	Red pigmented	Gram negative rod	Nil

Table 7: Gram's staining result for cultures obtained from Potiskum soil sample

s/n	Colony type	Surface soil	10cm depth soil
1	Spherical	Gram negative rod	Gram positive cocci
2	Irregular	Gram positive cocci	Gram positive rod
3	Yellow pigmented	Gram negative rod	Gram negative rod
4	Pink pigmented	Gram positive cocci	Gram positive cocci
5	white pigmented	Nil	Gram positive rod

Table 8: Gram's staining result for cultures obtained from Nguru soil sample

s/n	Colony type	Surface soil	10cm depth soil
1	Spherical	Gram Positive cocci	Gram Positive cocci
2	Irregular	Gram Positive rod	Gram Positive cocci
3	Rhizoid	Gram Positive rod	Gram Positive rod
4	Yellow pigmented	Gram Positive cocci	Gram Negative rod
5	white pigmented	GramPositivecocci	Gram Negative rod

Table 9: Gram's staining result for cultures obtained from Gashua soil sample

s/n	Colony type	Surface soil	10cm depth soil
1	Spherical	Gram positive cocci	Gram positive cocci
2	Irregular	Gram positive rod	Gram positive rod
3	Rhizoid	Gram positive rod	Gram positive rod
4	Yellow pigmented	Gram positive cocci	Gram positive cocci
5	Pink pigmented	Gram positive cocci	Gram positive cocci
6	Red pigmented	Gram negative rod	Gram negative rod

4.1.3. Observation for Application of Oil Utilizing Bacteria

After the test for the application of oil utilizing microorganisms, the following results were obtained:

Table 10: Cultures isolated from Damaturu soil showing growth on petroleum

s/n	Colony type	Growth on petroleum
1	Spherical shaped, gram positive cocci	–
2	Irregular shaped, gram positive cocci	–
3	Rhizoid shaped, gram positive rod	+
4	Yellow pigmented, gram positive cocci	–
5	Pink pigmented, gram positive cocci	–
6	Red pigmented, gram negative rod	–

Table 11: Cultures isolated from Potiskum soil showing growth on petroleum

s/n	Colony type	Growth on petroleum
1	Spherical shaped, gram positive cocci	–
2	Irregular shaped, gram positive cocci	–
3	Rhizoid shaped, gram positive rod	–
4	Yellow pigmented, gram positive cocci	–
5	Pink pigmented, gram positive cocci	–
6	Yellow pigmented, gram negative rod	+
7	White pigmented, gram positive rod	–

Table 12: Cultures isolated from Nguru soil showing growth on petroleum

s/n	Colony type	Growth on petroleum
1	Spherical shaped, gram positive cocci	–
2	Irregular shaped, gram positive cocci	–
3	Rhizoid shaped, gram positive rod	+
4	Yellow pigmented, gram positive cocci	–
5	Yellow pigmented, gram negative rod	+
6	White pigmented, gram positive rod	–

Table 13: Cultures isolated from Gashua soil showing growth on petroleum

s/n	Colony type	Growth on petroleum
1	Spherical shaped, gram positive cocci	–
2	Irregular shaped, gram positive cocci	–
3	Rhizoid shaped, gram positive rod	+
4	Yellow pigmented, gram positive cocci	–
5	Pink pigmented, gram positive cocci	–
6	Red pigmented, gram negative rod	–

4.2. Discussion

A microbial prospecting method for hydrocarbon research and exploration has been in use for the last five decades. It is an exploration method based on the premise that the light gaseous hydrocarbons migrate upward from subsurface petroleum accumulations by diffusion and effusion, and are utilized by a variety of microorganisms present in the sub-soil ecosystem (Rasheed et al., 2008). This research however, reveals that those oil degrading microorganisms can be isolated from oil contaminated sites, which is similar to the report of Gupte et al., (2015). Other researchers also reported that oil degrading microorganisms can be isolated from oil contaminated sites, but in their case they isolated microorganisms that degrade used oil (Khan et al., 2011).

Among the mixed cultures obtained in our present research study, only rhizoid shaped, gram positive rod culture isolated from soils of Damaturu, Nguru and Gashua continues growing in petroleum after 48 hours of incubation. This is the same as that reported by Khan et al., (2011) and Gupte et al., (2015). Also, among the mixed cultures obtained from soil of Potiskum and that obtained from 10cm depth soil of Nguru, a yellow pigmented, gram negative rod shows continuous growth in petroleum after 48 hours of incubation. This is also the same as reported by Gupte et al., (2015) where they isolated 12 gram negative rods hydrocarbon utilizing microorganisms. However, in this study the identification of bacterial strains were not carried out. The ability of those cultures to continue growing in petroleum shows the petroleum utilization ability of the cultures. This implies that oil-degrading, as well as oil indicating bacteria can be isolated from contaminated sites from major areas of Yobe State.

CHAPTER FIVE CONCLUSION AND RECOMMENDATIONS

5.1. Conclusion

Oil degrading microorganisms can be isolated from soils of major areas of Yobe State. The ability of these microorganisms to degrade oil shows their potentiality in detecting the presence of oil in an area. Their abundance in the soil, mayact as hydrocarbon signature on the soils and hence they may be considered as oil indicator microbes.

5.2. Recommendations

The following recommendations are made based on the findings of this research:

1. The geo-microbial methods for petroleum exploration can be used to detect the presence of oil in an area. Therefore, this work can be used to explore oil in Northern part of Nigeria.

2. Geologist can integrate this method of oil exploration with other geophysical and geochemical methods to prioritize drilling locations thereby reducing drilling risks and achieving higher success in petroleum exploration.
3. Researchers can use this work as a reference point to carry out further research on the same or relevant topic.

REFERENCES

Atlas, R.M. (1984). *Petroleum Microbiology.* New York: McMillan Company.

Adati, A.K. (2012). Oil exploration and Spillage in the Niger Delta of Nigeria. *Journal of Civil and Environmental Research*, 2(3).ISSN 2222-1719.

Beghtel, F.W., Hitzman, D.O., and Sundberg, K.R. (1987).Microbial Oil Survey Technique (MOST); Evaluation of New Field Wildcat Wells in Kansas. *Association of Petroleum Geochemical Explorationists Bulletin*, 3, 1-14.

Brisbane, P.G. and Ladd, J.N. (1965).The Role of Microorganisms in Petroleum Exploration. *Annual Review Microbiology.*19: 351-364.

Davis, J.B. (1967). *Petroleum Microbiology.* Amsterdam: Elsevier publishing company.

Dionisi, H.M., Harms, G., Layton, A.C., Gregory, I.R., Parker, J., Hawkins,S.A., Robinson, K.G., and sayler, G.S. (2003). Power Analysis for Real Time PCR Quantification of Genes in Activated Sludgeand Analysis of the Variability Introduced By D.N.A extraction. *Appl. Environ. Microbiol.*, 6597-6604.

Fan, Z., Yuehui, S., Yong, Z., Zhifeng, Z., Shuqiong, K. and Dujie, H. (2010).Molecular Biology Techniques Applied to the Microbial Prospecting of Oil and Gas in the Ban 876 Gasand Oil Field in China. *AppliedMicrobiology and Biotechnology,* 86, 1183-1194. Doi 10.1007/00253-009.2426-5.

Gupte, A., and Sonawdekar, S. (2015). Study of Oil Degrading Bacteria Isolatedfrom Oil Contaminated Sites. *International Journal for Research in Applied Science and Enginereering Technology.* 3(2): 345-347.

Hanson, R.S. and Hanson, T.E. (1996). MethanotrophicBacteria. *Microbiology Review*, 60, 439-471.

Hanson, R.S. and Wattenberg, E.V.(1991).Ecology of Methylotrophic Bacteria. In I. Goldenberg and J.S. Rokem, eds., *Biology of Methylotrophs* (pp. 325-349) London:Butterworths.

He, J.Z., Shen, P.J. Zhang, L.M., Zhu, Y.G., Zheng, Y.M., Xu, M.G., Di, H.J.(2007). Quantitative Analysis of the Abundance and Composition of Ammonia-oxidizing Bacteria and Ammonia Oxidizing Archeaof A Chinese Upland Red Soil Under Long-term Fertilization Practices. *Environmental Microbiology*, 9, 2364-2374

Hitzman, D.C., Tucker, J.D., Heppard, P.D. (1994). Offshore Trinidad SurveyIdentifies Hydrocarbon Microseepage: *26th Annual Offshore Technology Conference*, OTC 7378, Houston Texas.

Hugenholtz, P., Goebel, B.M. and Pace, N.R. (1998).Impact of CulturedIndependent Studies on Emerging Phylogenetic View of Bacteria Diversity.*Journal of Bacteriology*, 180, 4765-4774.

Kaster, K.M., Bonaunet, K., Berland, H., Kjeiler-Eilertsen, G., Brakstad, O.G.(2009). Characterization of Cultured-independent and Dependent Microbial Communities in A High-temperature Offshore Chalk Petroleum Reservoir. *Antonie Van Leeuwenhock*. 96:423-439.

Khan, J.A., and Abbas, S.H. (2011). Isolation and Characterization of Oil Microorganisms from Oil Contaminated Sites. *Advances in applied science research*. 2(3):455-457.

Leadbetter, E.R. and Foster, J.W. (1958).Studies on Some Methane UtilizingBacteria.*Archives Microbiology*. 30, 91-118.

Lopez, P.J., Tucker, J.D. and Hitzman, D.C. (1983). Hydrocarbon Microseepage Signatures of Seismic Structures Identified by Microbial Surveys, Sub-Andean region, Bolvia, *AAPG Annual convention program,*p.162.

Lysnes, K., Bodtker, G., Torsvik, T., Bjornested, E.S. (2009). Microbial Response to Reinjection of Produced Water in An Oil Reservoir. *Appl. Environ. Microbial,* 83,1143-1157.

Miller, G.H. (1976). Microbial Survey Helps to Evaluate Oil and Gas. *Oil andGas Journal*, 4, 192.

Mogilewskii, G.A. (1938). Microbiological Investigation in Connecting withGas Surveying*: Resvedka Nedr*, 8, 59-68.

Pareja, L. (1994). Combined Microbial, Seismic Surveys Predict Oil and Gas Occurrences in Bolvia.*Oil and Gas journal*. 24: 68-70.

Paula, B.M., Fernando, D.A., Armando, C.F., Justo, C.F., Dos Santos, E.V. andDe Olivera, V.M. (2011). Cultivation-independent Methods Applied to the Microbial Prospection of Oil and Gas in Soil from A Sedimentary Basin in Brazil.

Rasheed, M.A., Patil, D.J., and Dayal, A.M. (2013).*Microbial Techniques forHydrocarbon Exploration.*Open access chapter. Chapter 9, pp. 195-203.

Sealy, J.R. (1974a).A Geomicrobial Method of Prospecting for Oil.*Oil and GasJournal*, 8, 142-146.

Sealy, J.R. (1974b).A Geomicrobial Method of Prospecting for Oil. *Oil andGas Journal*, 15, 98-102.

Skovhus, T.L., Ramsing, N.B., Holmstrom, C., Kjelleberg, S., Dahllof, I.(2004). Real-time Quantitative P.C.R for Assessment of Abundance of Pseudo Alferomonas Species in Marine Samples.*Appl. Environ. Microbiol.*, 70, 2373-2382.

Tedesko, S.A.(1995). *Surface Geochemistry in Petroleum Exploration.* NewYork: Chapman and Hall, inc., 1-206.

Tucker, J. and Hitzman, D.C. (1994). Detailed Microbial Surveys Help Improve Reservoir Characterization: *Oil and Gas journal*, 92(23), 65-68.

Wagner, M. Piske, j. and Smit, R. (2002).Case Histories of Microbial Prospection for Oil and Gas.*AAPG Studies in Geology 48 and SEG Geophysical References Series*, vol. 11, pp. 453-479.

Whittenbury, R., Phillips, K.C., and Wilkinson, J.G. (1970). Enrichment, Isolation and Some Properties of Methane Utilizing Bacteria.*Journal of General Microbiology,* vol. 61, 205-218.